FIND YOUR NEXT JOB WITH CHATGPT
A GUIDE FOR EVERYONE

BY BEN GOLD

Copyright © 2023 by Ben Gold

FIND YOUR NEXT JOB WITH CHATGPT

All rights reserved. No part of this publication may be reproduced, distributed, or transmitted in any form or by any means, including photocopying, recording, or other electronic or mechanical methods, without the prior written permission of the publisher, except in the case of brief quotations embodied in critical reviews and certain other noncommercial uses permitted by copyright law. For permission requests, write to the publisher, addressed "Attention: Permissions Coordinator," at info@beyondpublishing.net

Quantity sales and special discounts are available on quantity purchases by corporations, associations, and others. For details, contact the publisher at the address above.

Orders by U.S. trade bookstores and wholesalers. Email info@BeyondPublishing.net

The Beyond Publishing Speakers Bureau can bring authors to your live event. For more information or to book an event contact the Beyond Publishing Speakers Bureau speak@BeyondPublishing.net

The Author can be reached directly at BeyondPublishing.net
Manufactured and printed in the United States of America distributed globally by BeyondPublishing.net

New York | Los Angeles | London | Sydney

ISBN Hardcover: 978-1-63792-639-0

ISBN Softcover: 978-1-63792-614-7

TABLE OF CONTENTS

ACKNOWLEDGMENTS	1
MY JOURNEY:	2
CUSTOMER TESTIMONIALS:	4
NAVIGATING THIS GUIDE	5

CHAPTER 1: ARTIFICIAL INTELLIGENCE 101 — 6

What is Artificial Intelligence?	7
Generative AI—The New Revolution	8
AI in the Job Market	9
Glossary of important AI Terms	11
Getting Started with ChatGPT	12
• GPT Paid vs. Free versions:	17
Input #1—Prompts	19
• Understanding the Limitations of Prompts	19
Crafting your first prompt	21
Tips for Crafting Effective Prompts	23
• Failed Prompting—an Example	24
Input #2—Contextual Data	25

CHAPTER 2: USING AI FOR INTERNAL PREPARATION — 27

Understanding AI usefulness in Career Search Phases	27
What career path is right for me? How to establish career objectives:?	30
Using AI to Compare Career Paths	34
Crafting Your Resume with AI	38
• Resume Section: Headline	39
• Resume Section: Core Proficiencies / Skills	42
• Resume Section: Accomplishments	44
• Resume Section: Professional Experience/Other Sections	46

Using AI to Improve your LinkedIn profile	48
• LinkedIn Section: Headline	49
• LinkedIn Section: About Me	50
• LinkedIn Section: Experience	52
• LinkedIn Section: Skills	52
• Using AI to become an expert in your role and industry	53

CHAPTER 3: LEVERAGING AI FOR OUTREACH — 58

Job Matching	58
Applying to Jobs	61
• Using AI for understanding compatibility	61
• Using ChatGPT to get Compatibility ratings	63
• Cover Letters in 10 Minutes or Less	65
• Using AI for Networking	68
Ways to interact with people on LinkedIn	74
#1 Engage with a post from someone of interest.	74
#2 Writing Connection Requests Where They Will Accept—300-character limit	75
#3 - Write a LinkedIn InMail or Email	78
Using AI for Job Interviews	80
Negotiate Your Way to Salary Success	88
The one profession ChatGPT will never replace: Comedians	91
Take Action Today and Become an AI Master to Propel Your Job Search	94
Your 60-Day Action Plan	95

Acknowledgments

I would like to thank my wife, Maria, for the encouragement to go out on my own and for her insightful and amazing suggestions for creating a tech startup and launching an AI coaching career.

I would also like to thank my business partners, Mihaela and Darren Warner, who were willing to start the journey with us within a week of me pitching the website idea that became JobCo. They have built out the web infrastructure, product functionality and APIs, but more importantly, have been fantastic counterparts during the creation and expansion of this business.

Finally, I thank Dan Seyler, Andrea Harvey and Georgiana Iancu for their instrumental mentoring as our team navigates the complexities of building this AI career coaching start-up.

My Journey:

January 26, 2023, marked a turning point in my life. I was let go from my Sales Engineering job at a technology company, and I found myself jobless for the first time in a decade. Because I spent 20 years in the technology field, the last 5 of which were specifically focused on AI solutions, I was able to use my experience to explore a recent AI innovation, ChatGPT, as a tool for personalizing my job search. I recognized generative AI as a game changer on a global scale.

I learned how to use ChatGPT to optimize my resume, understand compatibility with new opportunities, write cover letters and prepare for interviews. I noticed that the startup companies were the ones that were truly innovating AI and that there is a gap in knowledge in the job seeker market on how this technology will revolutionize the job market and career advancement.

I first decided to create an AI-powered career coaching startup that would disrupt the job-seeker market, and in understanding the issues job-seekers face, I have spoken with hundreds of recruiters, career counselors and job-seekers. I decided to spread the message through webinars, personal coaching, and this book.

AI has revolutionized how we interact with the world and each other. We've all experienced it. Shaping our social interactions, information consumption, and spending habits is only the beginning. You're about to learn how it's also transformed how we enter, pivot, or accelerate our career paths. I do not promise that this book will make you a millionaire overnight, nor will you find a job after applying all the principles. I promise you will get honest feedback on whether your skills are still in demand and receive suggestions on other paths you can choose to succeed professionally.

My mission is to convey the lasting relevance of AI to job seekers and equip them with accessible tools. This enables individuals—

regardless of their prior AI knowledge—to learn, master, and leverage AI to secure their next job and enhance their present professional effectiveness.

Ben Gold
August 16, 2023

Customer Testimonials:

"It's an absolute must-read for anyone in the market for a new career or for advancement in their current one. I can't stress strongly enough how simple my job search became after using the tools in this book. "Life-changing" is the best description I can muster. The instructions and descriptions are straight-forward and clear. Organized in an intuitive and user-friendly manner, this book walked me through simple activities anyone can complete, and I mastered the concepts immediately."

Beth F

"Find your next Job with ChatGPT was a game changer for me as a job seeker.

The techniques and tips shared in the book for leveraging AI in the search process were incredibly eye-opening. Practical strategies for tailoring my resume, crafting personalized cover letters, and even simulating interview scenarios with ChatGPT gave me a competitive edge.

Thanks to this resourceful guide, I have developed a newfound confidence in navigating the modern job market. A must-read for anyone looking to stand out and succeed in their job search."

Sally S

Navigating This Guide

Over a span of six months, I've immersed myself in research and hands-on experimentation with many strategies to help job seekers harness the power of the newest technologies. Regardless of whether you're a tech enthusiast who leverages Artificial Intelligence models regularly or someone who still associates Artificial Intelligence with the *Terminator* movie, this guide promises to be a repository of valuable insights that will aid in charting your career path, refining your resume, readying you for interviews, and dissecting your resulting job offers.

In creating this guide, I've interacted with hundreds of job aspirants, recruiters, employers, and career consultants to understand current employment trends. This guide is a pragmatic blueprint for understanding these trends and acknowledging AI as not just a fleeting trend but a transformative force with enduring implications across every major industry.

This guide unfolds in three primary sections:

Chapter 1—Artificial Intelligence 101 This is for people with less experience with AI. It outlines why AI is so significant and offers practical advice on getting started.

Chapter 2—Harnessing AI for Internal Preparation: This section focuses on how you can employ AI to shape your career objectives and optimize your resume and LinkedIn profile.

Chapter 3—Leveraging AI for Outreach: The final segment sheds light on outreach activities like networking, job applications, cover letters, interviews, and salary negotiations.

Chapter 1: Artificial Intelligence 101

People who Don't Use AI

People who Use AI

People who Master AI

What is Artificial Intelligence?

Artificial Intelligence (AI) is a branch of computer science concerned with creating and developing intelligent machines capable of performing tasks that would typically require human intelligence. These tasks include speech recognition, decision-making, visual perception, language translation, etc.

AI has been an important tool used by large corporations for some time. How many of the following do you recognize?

- **Google Search Algorithms:** Google has been using AI in its searches for years, including algorithms like RankBrain, a machine learning-based algorithm, to provide more relevant search results.

- **Amazon's Recommendation System:** Amazon has been using AI to recommend products based on customer browsing and purchasing history. The AI learns from customer behavior to suggest items they are likely to be interested in.

- **Netflix's Personalization Algorithm:** Netflix uses AI to offer personalized recommendations to its subscribers. The AI learns from a user's viewing history and preferences and suggests TV shows and movies they might enjoy.

- **Spam Filtering:** Email service providers like Google's Gmail and Microsoft's Outlook have used AI for years to filter out spam emails and protect users from malicious threats.

- **Facebook's News Feed:** Facebook uses AI to decide what content to show in each user's news feed based on their preferences, interactions, and behavior on the platform.

- **Fraud Detection:** Banks and credit card companies have been using AI to detect unusual patterns of activity that may indicate fraud.

- **Voice Assistants:** Apple's Siri, Google's Assistant, and Amazon's Alexa use AI to understand and respond to voice commands.

Generative AI—The New Revolution

Generative AI is a subset of AI that takes things a step further. Instead of just learning from data and making decisions or predictions based on that data, generative AI models can create new data that is similar to or a continuation of, the data it was trained on. This is done by understanding the underlying patterns or distribution of the input data and generating new data that follow these same patterns.

For example, generative AI is the technology behind creating realistic images, writing human-like text, composing music, or even creating new video game levels. It's also what powers the advanced language models like ChatGPT by OpenAI, which can generate human-like text based on the prompts given to it. While AI involves systems that can learn and make decisions, generative AI systems can also generate new, original content.

> Quick Note: Language models are powerful tools but have limitations. They can often generate grammatically correct but semantically nonsensical text, as they lack a true understanding of the world and the text they generate. They can also replicate and amplify the biases in the data they were trained on.
>
> Furthermore, language models can be unpredictable. A language model can produce vastly different results with the same prompt twice. This is because many models, especially large ones like GPT-4, inject randomness into their outputs to generate more diverse responses.

AI in the Job Market

By creating an intuitive tool that can cut certain manual tasks such as writing and coding by 5-10x, AI is transforming how we find and keep jobs. As a job seeker, it is important to use AI to strategize the best options for gaining employment and also to understand how AI is changing the industry they are targeting.

Employers are using AI to screen candidates and evaluate employee performance based on how well employees can implement AI to be more efficient and effective at their jobs.

> An Applicant Tracking System (ATS) is a software application that automates the recruitment and hiring process, enabling organizations to collect, sort, and filter resumes and applications. By ranking candidates based on specific criteria and keywords, it assists recruiters and hiring managers in identifying the most suitable candidates for a job opening, thus streamlining the entire process.

Here are examples of how generative AI is transforming several industries:

- **Marketing:** Generative AI produces personalized advertising content, dynamically generating text and visuals that resonate with individual consumers. It also helps create numerous variations of ads for A/B testing.
- **Real Estate:** Generative AI can aid property valuation by generating estimates based on large parameters. It can also be used in architectural design, producing multiple design variations based on the specified criteria.
- **Technology:** In the tech industry, generative AI is used in software development to automatically generate code, thereby speeding up the development process. It's also used in designing new AI models, a process known as AutoML.

- **Finance:** In the financial sector, generative AI can simulate various market scenarios to aid risk assessment and investment strategy planning. It can also generate financial reports or conduct analysis based on financial data, providing valuable insights for decision-making.
- **Healthcare:** Generative AI can create synthetic medical images for AI model training and generate novel drug candidates, enhancing disease detection and drug discovery.
- **Sales:** Generative AI can produce personalized sales pitches and simulate training scenarios. Additionally, it aids in forecasting sales based on multiple factors.

AI is also transforming existing jobs by automating routine tasks such as answering basic customer service questions, scheduling, and data entry and augmenting more complex ones involving strategic decision-making or predicting market trends.

By automating routine and repetitive tasks, AI allows human workers to focus more on strategic, creative, and decision-making tasks that add more value to their roles; however, this will also make it imperative for each person to learn how to use AI in their current profession.

Glossary of important AI Terms

AI (Artificial Intelligence): Technology designed to mimic human intelligence and behavior.

Generative AI: A type of AI that creates new, original content based on its learning, such as GPT.

ML (Machine Learning): A type of AI that enables a system to learn from data autonomously.

Natural Language Processing (NLP): A field of AI that enables machines to understand, interpret, and generate human language.

Input: Data you feed into the AI model is broken into two categories:

- **Prompt**: The commands given to an AI model that guides its response.
- **Contextual Data**: Information relevant to a specific situation that an AI model uses to form appropriate responses. Examples include Resumes, Job Descriptions, and Social Media profiles.

Output: The response generated from your input.

OpenAI: An artificial intelligence research lab that created ChatGPT.

GPT (Generative Pre-Trained Transformer): A type of AI language model developed by OpenAI.

Token: In language models, a unit of text that the model reads. It could be as short as one character or as long as one word.

Getting Started with ChatGPT

(If you are already using GPT 4.0, you can skip this section.)

Go to https://openai.com/blog/chatgpt, click on "sign up," and create an account.

Figure 1: OpenAI's Blog Introducing ChatGPT[1]

You can use an existing Google, Microsoft, Apple Account or an email address.

---

[1] https://openai.com/blog/chatgpt

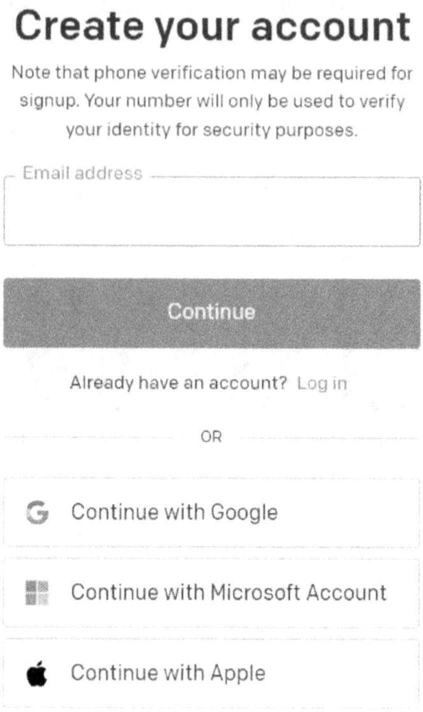

Create your account

Note that phone verification may be required for signup. Your number will only be used to verify your identity for security purposes.

Email address

[Continue]

Already have an account? Log in

──── OR ────

G Continue with Google

▦ Continue with Microsoft Account

  Continue with Apple

Figure 2: OpenAI's Sign-up Page[2]

After successfully creating your account, you will be asked to choose between ChatGPT, DALL-E or API. DALL-E is Open AI's image creation application, while the API area is how companies can integrate their websites and leverage ChatGPT capabilities.

---

[2] https://shorturl.at/OV178

Figure 3: OpenAI Apps[3]

After you click the ChatGPT box, you will notice a search bar with the text "Send a Message" at the bottom of the page. This is where all prompts and contextual data are added to get GPT output. Notice on the top that this is an example of a paid version. The program defaults to the GPT 3.5 version. For each chat, you need to manually click the GPT-4 button to use the GPT-4 version.

---

[3] https://platform.openai.com/apps

Figure 4: ChatGPT Main Screen[4]

On the left of the screen, ChatGPT remembers every individual chat or conversation that you have had. The best practice is to create a new chat whenever you want to begin a new subject.

> Advanced Tip: You can always return to older chats by clicking on the specific conversation. As you apply to different jobs, the best practice is to label each chat by opportunity. This way, it is much easier to leverage the contextual data you created (for example, from the job description and your resume) when the opportunity moves to the interview phase.

---

[4] https://chat.openai.com/

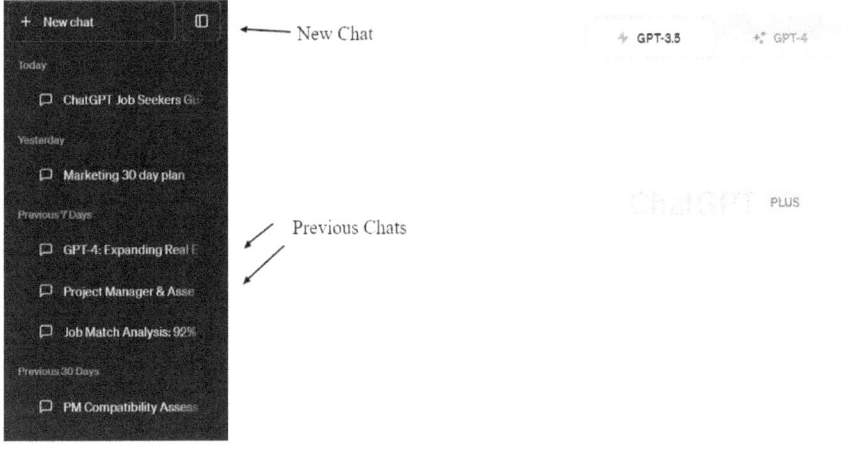

Figure 5: Sidebar of ChatGPT[5]

---

[5] https://chat.openai.com/

GPT Paid vs. Free versions:

If you are brand new to ChatGPT, you can definitely take your time to familiarize yourself with how the AI functions. You can test many of the prompts in this book and see the output. The $20/month investment should be a no-brainer.

Here are the differences:

GPT 3.5 does offer faster output than 4.0 and is free. However, GPT 4 is trained on 10x the parameters as GPT 3.5.

- GPT 4.0 is smarter and more perceptive. Cover Letters, Compatibility analysis and interviewer insights are far better in the 4.0 version.
- It has a larger memory. The GPT versions use a concept as tokens (1000 tokens = about 750 words) to measure the amount of text it can evaluate and generate output. When you have a very long "conversation" with the AI, it will "forget" or no longer incorporate older data in its analysis.

GPT 3.5 can retain 4000 tokens, while 4.0 can retain 8000 tokens. This means that if we ask the AI to consider a resume, job description, and social media profile to generate anticipated questions and the effectiveness of interview responses, GPT 4.0 will be able to crunch this data set much better than older versions.

- GPT 4.0 is multilingual, i.e., it can accept inputs in multiple languages

- GPT 4.0, as of this writing, has a limit of 25 prompts every 3 hours.

AI HUMOR: The one question ChatGPT 4.0 cannot answer:

PROMPT: What is the difference between GPT 3.5 and GPT 4.0?

ChatGPT RESPONSE:

As of my knowledge cutoff in September 2021, OpenAI has not released any models beyond GPT-3. Therefore, I can't provide specific differences between GPT-3.5 and GPT-4.0.

Generally, new versions of models like GPT tend to focus on improving performance, reducing bias, increasing the understanding of nuanced language, and refining the model's ability to generate text that is more contextually aware and accurate.

For any further updates or specific changes in the model after GPT-3, please refer to the most recent information available from OpenAI or other reliable sources.

Input #1—Prompts

A prompt is a command a user creates to generate a specific output from the AI model.

Prompts play a crucial role in determining the output of GPT models. The information contained in the prompt guides the model's text generation, influencing not just the topic but also the tone, style, and sometimes even the format of the generated text. People who have mastered this process are referred to as Prompt Engineers.

For instance, if you give the AI model the prompt, "Translate the following English text to French: 'Hello, how are you?'", the model understands from the prompt that it needs to translate the text into French. Alternatively, if you input a more complex prompt like "Write a short, spooky story that starts in a haunted mansion," the model will generate a narrative fitting your specifications.

Understanding the Limitations of Prompts

While prompts are a powerful tool for guiding a GPT model's output, they have limitations. A prompt is only as good as the contextual data you supply. If a model doesn't have relevant data, it may struggle to generate a useful response, no matter how well-crafted the prompt.

For example, if you don't take the time to understand and document your career goals, skills and achievements, the AI response will not yield insights that can give you actionable data.

IMPORTANT NOTE: Prompts can't guarantee accuracy. Even with a perfect prompt, GPT models might produce plausible-sounding but incorrect or nonsensical outputs. It's crucial to bear these limitations in mind when interpreting and using the output of GPT models.

A FOLLOW-UP BEST PRACTICE: NEVER SEND AI OUTPUT TO AN EMPLOYER IN THE FORM OF A COVER LETTER OR EMAIL WITHOUT CAREFULLY READING THE TEXT AND VERIFYING IT TRULY EXPRESSES YOUR BEST INTERESTS.

Crafting your first prompt

> NOTE: If you have not yet set up your ChatGPT account, now is the time to put this book down and register. By doing the four examples below, you can get oriented on how generative AI works.

As a new user, the best way to understand how prompts work is to start simply.

PROMPT: Give me a 5-day itinerary for Costa Rica (or any other destination)

You will notice that the AI will give you a generic response of a 5-day trip.

Let's add some more context:

PROMPT: I am traveling with my wife and two kids.

The AI will know what you just said refers to the previous prompt. There is no need to repeat the whole context. Notice how the itinerary has changed to focus on family activities.

We can get more specific:

PROMPT: 3 days of birdwatching (or any other activity)

The AI understands a family is on a 5-day trip and wants 3 days to include birdwatching activities.

Finally, let's dig deeper on Day 2:

PROMPT: Expand on day 2 activities

As you begin interacting with ChatGPT, learning how to get big-picture responses and dig deeper into items you want to explore is important.

Congratulations! You have started on the journey of prompt mastery.

Tips for Crafting Effective Prompts

As you can see, prompts play an essential role in guiding a language model like ChatGPT to generate the desired output. Crafting effective prompts enhances the utility of these models to a great extent. Here are some tips for creating useful prompts:

- **Be Clear and Specific**: Precision and clarity are crucial when crafting prompts. The more specific your prompt, the more targeted the response will be. For instance, instead of "Write a story," you could state, "Write a short mystery story set in Victorian England involving a stolen artifact."
- **Guide the Tone and Style**: The language and tone of your prompt can help guide the model to generate text in a similar style. For instance, if you want a more formal output, use a formal tone in your prompt. On the other hand, if you're looking for a creative, casual response, use a more casual tone.
- **Use Instructional Language:** Instructional prompts can be very effective, especially when you need the model to perform a specific task. For example, if you want a summary of a lengthy article, your prompt could be, "Summarize the following article in 3 concise sentences."
- **Set the Context**: If the task requires understanding a certain context, provide that context in the prompt. For example, if you're asking for a summary of a complex topic, you could start with a sentence or two about the topic to set the stage.
- **Experiment and Iterate**: It's important to remember prompt crafting can involve trial and error. Don't be afraid to experiment with different approaches and iterate based on the results you get. Over time, you'll get a better feel for how to craft prompts that get the results you're looking for.
- **Use Roles:** In a conversation with the AI model, you can use roles at the beginning to gently instruct the assistant about the behavior you want. For example, a message like, "You are an assistant that speaks

like Shakespeare," can guide the model's tone throughout the conversation.

Failed Prompting—an Example

What happens when you feed data without the proper context or commands? Here is an example. I wanted to write an email to my seminar students and invite them for a coaching session, but I forgot to add the command "Improve this Text." Here is the full conversation:

PROMPT: I appreciate your participation in yesterday's discussion "AI Powered Job Searching Strategies." As a follow up, I am offering everyone who has volunteered for the beta test a one-hour coaching session to understand your specific needs and how AI can be utilized in your job search. Feel free to click on the link and book the session.

I look forward to working with you.

Ben

ChatGPT RESPONSE:

Hello Ben,

Thank you for the invitation. However, as an artificial intelligence, I do not have personal needs or a career to pursue. Therefore, I do not need to book a coaching session. But I am more than willing to help in any way I can by providing information, discussing AI strategies or answering any questions you may have.

Best,

ChatGPT

Input #2—Contextual Data

In AI and machine learning, contextual data refers to additional information that helps a system understand the circumstances or environment around a specific data point or an event. It's the auxiliary information that adds meaning to the primary data and allows a more nuanced interpretation of it.

Contextual data is critical for several reasons:

- Enhanced Understanding
- Improved Predictive Capabilities
- Reduced Ambiguity

Contextual data can come in many forms; however, in the context of a job search, there are 3 primary elements you'll use

- Resume
- Job Description
- LinkedIn profile of the person(s) conducting an interview

Because the resume is contextual data produced 100% by you, that's evaluated by the AI (and employers who receive your application), this is where you'll need to focus a large amount of energy at the start of your job search.

Contextual data examples from your resume:

- **Career Progression and Experience:** Your employment history, which includes the roles you've held, the tasks you've undertaken, and the organizations you've worked for, provides context about your career progression and practical experience. It shows potential employers what you've accomplished and how you might fit into their organization.
- **Skills and Qualifications:** The skills and qualifications you list on your resume give context about your abilities and the tasks you're equipped to handle. This might include technical skills, like

programming languages, or soft skills, like communication or leadership.

- **Education and Certifications:** Your educational background provides context about your foundational knowledge and areas of formal study. Certifications or additional training can add further context, showing areas where you've sought to expand your skills and expertise.
- **Personal Projects and Volunteer Work:** Personal projects, volunteer work, or other extracurricular activities can provide additional context about your interests and skills that might not be evident from your formal employment or education history.

Chapter 2: Using AI for Internal Preparation

> Leverage career counselors and resources to identify passions, refine your resume, and enhance your LinkedIn profile. The accuracy of AI models improves when you detail your interests and achievements.
>
> While self-guided exercises can be helpful, they don't replace the guidance of a career mentor or a job-seeker community. If you're working with a career advisor, AI tools can complement their services by streamlining activities and enhancing documentation. The list below shares in more detail areas where AI is more and less effective.

Understanding AI usefulness in Career Search Phases

Each phase is appraised on a scale of 1-5 based on the efficacy of AI in executing the tasks in that phase.

5 = **AI will be the primary tool used to execute the task.**

4 = **AI will be the primary tool combined with other elements.**

3 = **Use a mix of AI and other tools.**

2 =- **AI can help the process but start with other tools.**

1 = **ChatGPT, as of today, is not the best place to start or execute the task.**

Defining Career Objectives: 3

AI can assist in formulating data-driven suggestions. However, networking with other job **seekers** and consulting professionals current with the market trends is advised. Various networking groups provide standardized tests to help define personal goals and career objectives.

Updating your Resume: 3

With appropriate prompts, ChatGPT can assist in crafting the upper sections of your resume: headline, core competencies, and accomplishments. Collaborating with a resume expert is advised for final edits and formatting to ensure compatibility with ATS (Applicant Tracking System).

Updating your LinkedIn Profile: 3

ChatGPT can use your resume information to update your Headline and About sections while using your resume updates to enhance your experience section. Consulting a LinkedIn expert for a profile review is suggested to optimize your visibility to recruiters and hiring professionals.

Job Matching: 1

ChatGPT data only extends up to September 2021. Therefore, aspects such as company profiles, job vacancies, or industry trends need to be sourced by other methods. Even though there are plugins that allow more recent data to be evaluated by ChatGPT, I suggest using LinkedIn as a primary source of job leads as many job sites are filled with spam and databases like Crunchbase can aid in researching targeted companies. Once you have identified an opportunity that looks like a great fit, then AI can help you from application to salary negotiation.

Networking: 5

ChatGPT can help you understand the priorities and backgrounds of the people you want to network with. This helps you to craft a message more likely to get a response.

Applying for Jobs: 5

Once a job opening is identified, AI can help ascertain compatibility as a percentage, identify skills gaps, and suggest improvements to your resume. This allows for objectively evaluating job fit and how companies employing similar technology perceive your skillset.

Writing Cover Letters: 5

For applications where this is an option, ChatGPT can assist in crafting a personalized cover letter. Please refer to the cover letter section for potential pitfalls and best practices.

Interview Preparation: 5

AI can provide tailored preparation advice based on the interviewer's social media profile. This aids in predicting potential questions and gaining feedback on your responses.

Salary Negotiation: 4

Several factors should be considered before a salary offer is made. The AI can evaluate and guide you through the negotiation process by inputting the job title, location, and offer specifics. Because the model is based on data from 2021, though, industry standards might have changed.

What career path is right for me? How to establish career objectives:?

In this stage, it's important to spend considerable time soul-searching. Whether you are unhappy in your current role, just got laid off, or are graduating from school, determining your areas of competence and passion deserves attention.

> SUGGESTION: I highly recommend plugging into local job seeker groups and considering investing in a career advisor. This stage is vital to the rest of the career journey. If you don't know where you are going, AI can assist you in getting there.

The other reason this stage is so critical is that most of the contextual data you will feed into ChatGPT will be based on your work here.

Formal self-assessment tools are available. If you have taken one recently, you can enter the results into ChatGPT for further analysis. Examples of these kinds of tests include:

- Myers-Briggs Type Indicator (MBTI)
- StrengthsFinder (Now called CliftonStrengths):
- Strong Interest Inventory
- DISC Assessment
- Self-Directed Search (SDS)
- CareerExplorer

If you want to begin this process without a formal skills assessment test, you can use a technique designed to answer behavioral interview questions. This is important so early in the process because your career journey is a story of who you are, what you're good at, and what you love to do. Concrete stories of when you accomplished important goals under pressure are part of that story.

I will share one of these techniques known as **STAR**:

- S: Describe the Situation.
- T: What Tasks were involved?
- A: What Action did you take?
- R: What was the Result?

> SUGGESTION: Write down as many stories as you have in your professional life (or your personal life if the story illustrates a skill or passion transferable in your career). The goal is to come up with at least 6.

Here is an example from my previous role as a Solution Consultant:

Situation: The organization had a global sales force without any demo collateral that showed the company's global capabilities. Sales reps were using live client data to show business value.

Task: The challenge was to create a single global demo entity that highlighted the linguistic and localization capabilities.

Action: After evaluating the disconnected and illogical demo environments, I created a single streamlined organization that reflected situations the larger global clients faced. This included uploading documentation that resonated with UK, French, German, Australian, and Mexican audiences.

Results: The new GlobalCo demo environment was adopted by the whole organization and used to expand its global reach.

I am sharing a long response in this example to show how ChatGPT can take a small amount of information and provide a very insightful and detailed response.

PROMPT: What conclusions can you draw about my skill set, goals, and professional roles I should pursue based on this STAR analysis?

Chatgpt RESPONSE:

Skills:

- **Project Management and Organizational Skills:** Your ability to evaluate and restructure the demo environments signifies your strong project management and organizational capabilities.
- **Cross-cultural Understanding and Adaptation:** The successful creation of documents for diverse international audiences indicates a strong understanding of cross-cultural communication and the ability to adapt materials to fit different cultural contexts.
- **Technical Skills:** Your involvement in creating a demo environment suggests that you have good technical skills, possibly in software development, data management, or related fields.
- **Strategic Thinking and Problem Solving:** You were able to identify a major issue within the organization and devise a comprehensive solution that addresses not only the immediate problem, but also potential future issues, demonstrating strategic thinking and problem-solving capabilities.

Goals:

- **Process Optimization**: Your aim appears to be improving and streamlining processes within the organization, suggesting an interest in continuous improvement and efficiency.
- **Global Engagement:** You seem to have a goal of enhancing the global reach and impact of organizations, which can involve working closely with international teams, understanding global markets, and creating solutions that have wide appeal.

Professional Roles:

- **Product Manager or Project Manager:** Your demonstrated skills and goals align well with roles such as a Product Manager or Project Manager, where you would be in charge of overseeing the development and execution of projects, often with a global reach.
- **Global Strategy Consultant:** With your experience in global outreach and understanding of multicultural aspects, a role in global strategy consulting could be a good fit. You would assist organizations in understanding and entering new international markets.
- **Sales Enablement Manager:** Given your experience in creating demo environments and materials for sales teams, a role in Sales Enablement would be well-aligned. This could involve developing tools, content, and training to improve the productivity of sales teams.

As you get more comfortable with prompting, you will learn that it is OK to go "off script." Some AI answers might not resonate with you. Others might be intriguing. AI will give you better output by sharing more data with the model. As you find interesting career paths and professional goals, you'll have a starting point for more in-depth research.

Remember, at any point, you can ask the AI to dig deeper into any intriguing observations it offers.

Using AI to Compare Career Paths

What if you're not sure about your career path? You can get insightful answers once you've fed accurate data into ChatGPT. One of my students, who was debating whether he should pursue a career in sales or project management, received the following response to this prompt:

PROMPT: He is considering a sales or project management role. Based on the resume and other data provided, please explain the pros and cons of each.

ChatGPT RESPONSE:

Sales Role:

Pros:

● *Performance Rewards:* In many sales roles, earnings are closely tied to performance. This can provide a strong financial incentive and reward for hard work.

● *People Interaction:* Sales roles often involve a high level of interaction with clients and customers, which can be rewarding for people who enjoy interpersonal communication.

● *Autonomy:* Salespeople often have significant control over their schedules and may have the opportunity to work independently, managing their own clients and accounts.

Cons:

● *Income Uncertainty:* Because much of a salesperson's income is often based on commission, earnings can fluctuate widely and may be unpredictable.

● *Pressure:* Sales roles can be high-pressure, especially when targets or quotas aren't being met. This can lead to stress.

● *Rejection:* Sales roles often involve facing a lot of rejection, which can be difficult for some people to handle.

Project Management Role:

Pros:

- **Leadership Opportunities**: Project managers are often in leadership positions, responsible for guiding a team to achieve specific goals. This can be rewarding for people who enjoy leadership and decision-making roles.
- **Variety:** Project management can involve a wide variety of tasks and can provide opportunities to work on different types of projects, which can be engaging and challenging.
- **Transferable Skills:** The skills you acquire in project management, such as time management, organization, and team coordination, are highly transferable and can be valuable in a wide range of roles and industries.

Cons:

- **Stress:** Project management can be stressful, particularly when projects are complex, timelines are tight, or unexpected issues arise.
- **Responsibility:** The success or failure of a project often falls on the project manager's shoulders, which can be a lot of pressure.
- **Resource Constraints:** Project managers often have to make do with limited resources, including budget, staff, and time, which can be challenging.

PROMPT: Based on this resume, please give a compatibility rating as a % to him pursuing sales with 2 paragraph description and then repeat for project management

ChatGPT RESPONSE:

Sales Role Compatibility: 80%

His extensive experience in driving business growth, executing marketing strategies, and managing key accounts demonstrate strong *competencies for a sales role. His successful track record of strategizing, developing, and managing solutions that have addressed complex business challenges indicate a keen understanding of client needs and market trends. His skills in selling 6-figure training programs and taking a wellness business to seven figures in revenue clearly suggest his ability to close high-value deals. However, while his experience is impressive, the primary focus of his career has not been exclusively on direct sales, which may require additional training or transition time.*

Project Management Role Compatibility: 90%

He has showcased strong project management skills throughout his career. His experience in leading cross-departmental initiatives, overseeing teams, managing vendor contracts, and successfully implementing complex projects aligns strongly with the expectations of a project management role. His leadership role in co-founding and managing his startup demonstrates his ability to steer a project from conception to completion. The fact that he led multiple enterprise innovation initiatives that were utilized by top companies such as National Geographic, Atlantic Media, and AARP further emphasizes his project management skills. His extensive experience in this area gives him high compatibility for a project management role.

PROMPT: Are there any other professional paths where the compatibility is above 90%

ChatGPT RESPONSE:

Innovation Consulting: Compatibility 95%

His resume exhibits a high level of expertise in the field of innovation consulting. His diverse experience in strategizing, developing, and implementing innovative solutions to complex business challenges, as well as leading enterprise innovation initiatives, would make him an excellent candidate for senior roles in this field. As a co-founder and managing director, he has successfully created and implemented cross-departmental leadership development and mental wellness initiatives that surpassed financial goals. His understanding of emerging market trends, consumer insights, and technologies would be highly valuable in this role.

Leadership Development/Executive Coaching: Compatibility 90%

Another area where he might excel is in Leadership Development or Executive Coaching. His experience in developing innovative programming to guide and mentor executives at companies like Google, Microsoft, Salesforce and Adobe, shows his aptitude for coaching and guiding top-level professionals. The integration of values such as care, respect, optimism, and resilience into leadership and staff training programs demonstrates his deep understanding of organizational behavior and leadership principles. His success in this area could translate into a very effective career as an executive coach or leadership development consultant.

Crafting Your Resume with AI

Unless you're a student with no work experience, you'll likely have some form of a resume or CV that describes your past:

- Accomplishments
- Skills
- Jobs
- Education

When you are happily employed, there is no need to update your resume based on the latest trends in the marketplace. Many job seekers find themselves with a resume from several years back and are unsure of the best path to update. Here are some suggestions:

If your resume has not been professionally reviewed in the last 12 months, I highly suggest doing so. For example, The format widely used in 2019 is outdated in 2023.

> IMPORTANT: If your resume is not optimized to be read by ATS applications (Applicant Tracking System), it may be rejected on formatting only. Advances in AI technology on the employer side also make it imperative you are aware of employers' formatting and content expectations.

AI can also assist in updating your resume. While many formatting and presentation philosophies exist, I will share some simple techniques you can use to effectively update your resume, bringing it in line with current standards.

Step 1 – Identify a resource willing to review your resume

Step 2 – Ask for an example of a resume optimized for today's market

Step 3 – Use AI as illustrated in the following examples to update the content of your resume *before* speaking with the career resource:

Materials Needed:

1 – Most updated resume.

2 – Completed STAR content (related to career goals and accomplishments).

3 – Details of jobs held since the most recent resume update.

> Note: Because of the 8000 token memory limit, you might want to summarize the main points if the data in parts 2 or 3 is very detailed.

Resume Section: Headline

Headline Section: This is a brief introduction to your professional profile. It's a short statement or a title that appears at the top of your resume after your name and contact details. The purpose of a headline is to quickly grab the recruiter's attention by summarizing your professional identity in a concise, compelling manner. It should highlight your most relevant skills, experiences, or roles. This area typically has your preferred job title and 4-6 bullet points showcasing the main attributes you offer a company. It starts with the job title you aspire to (which should be similar or exact to the job title you are sending it in for).

Here is an example of my latest resume version before I decided to start my own business:

Senior Sales Engineer

- Accomplished software pre-sales professional with focus on AI and Machine learning and extensive experience in the Healthcare, Finance, Energy, Telecommunications, eCommerce, Construction, Manufacturing, Retail verticals.
- Excel at collaborating with cross functional teams such as sales, marketing, product and data science teams to ensure marketing to increase revenue and profits.
- Expertise in translating company product capabilities and innovations into business value and ROI for prospects/current clients.
- Proven ability to identify customer needs and provide innovative solutions that increase efficiency and improve operations.
- Possess strong technical knowledge and proficiency in APIs and integrations (Salesforce, Google Ads/Analytics, Adobe). Extensive global experience - willing to travel and or relocate internationally.

How do you get from having an outdated resume to having bullet points for your headline? I will share an example of a client I coached that had not updated his resume in 8 years. Based on the contextual data I suggested above (an old resume and updated examples of what he has done since the resume), we were able to create the following output:

PROMPT: You are a top-notch resume coach and must update the following resume. Make a summary of the main headline based on the following information (paste your resume):

ChatGPT RESPONSE:

• *Visionary innovation leader with over 15 years of experience strategizing and deploying solutions to address complex business challenges and stimulate growth.*

- Expert in analyzing market trends, generating consumer insights, and utilizing emerging technologies to create sustainable products and services.
- Proven competence in business strategy, project management, marketing, and community engagement.
- Known for leading enterprise innovation initiatives for major clients including National Geographic, Atlantic Media, and AARP.
- Co-founder of a successful wellness and leadership development enterprise, with a client roster including Google, Microsoft, Salesforce and Adobe..
- Accomplished mentor capable of integrating core values into leadership development and fostering a culture of resilience and optimism in teams.
- Skilled in the integration and support of SaaS products such as Slack, Basecamp, Google Workspace, and Salesforce.
- Demonstrated proficiency in business sales process transformation, responsible for taking a wellness business from mid-six figures to seven figures in revenue.
- Expertise in creating strategic partnerships and alliances to enhance business visibility and recognition.
- Credited with marketing an executive coaching product used by several Fortune 100 companies, including C-Suite executives at Google.

Having the AI present more than six options is fine. This exercise is not designed to give the final output. Instead, getting from random thoughts to several key elements will be a huge time saver. With this raw data, you can determine how to best use it. If you want to see different examples, you can ask the AI to refine the following:

PROMPT: Now take the resume and make the bullets geared toward a _____ position

Resume Section: Core Proficiencies / Skills

This resume section focuses on keywords you can use to pass through ATS keyword filters. The words you use should reflect both current skill sets and the skills you are looking to use in your future job. AI can help you determine the majority of terms needed.

Here is an example of my Core Proficiency section prior to starting my company:

CORE PROFICIENCIES

Artificial Intelligence | Machine Learning | Sales Enablement | Pre-Sales | Cross Functional Collaboration | NLP | New Business | Current Client Upsell | Competitive Selling | Global Sales | Team Leadership | (SaaS) | Integrations and APIs | Sales Demonstrations | Information Security | Innovation | Project Management | Quality Analysis | Coaching/Mentoring

Here is an example for one of my students:

PROMPT: Under the core proficiency section of the resume, please list the top 20 skills that summarize his abilities and experience:

ChatGPT RESPONSE:

Project Management

Product Ownership

Agile Methodologies

Business Analysis

Product Development

Stakeholder Management

Cost Saving Initiatives

Technical Operations

Process Improvement

Data Analysis and Management

Team Leadership and Training

Communication Skills

Problem-Solving

Attention to Detail

Regulatory Compliance

User Acceptance Testing (UAT)

Software Development Life Cycle (SDLC)

Jira and Confluence proficiency

Azure Databricks

Onboarding and Training

Resume Section: Accomplishments

The accomplishment section of a resume focuses on tangible achievements throughout your career. Rather than just listing duties or tasks performed in past roles, the accomplishment section illustrates the impact of your work. These could be quantifiable successes such as "Increased sales revenue by 50%" or notable milestones like "Led a project that won Industry Innovation Award." This section is essential to show recruiters what you've done and the positive results and value you've delivered.

Here is an example from my resume:

ACCOMPLISHMENTS

- Led alpha test collaboration with Data Science team and Product on new AI algorithm by creating a full documentation game plan that identified the best alpha test clients for low risk participation and feedback to be incorporated in the new product.
- Coordinated response and led internal communication and of 500+ questions with answers from product, security, sales, and CSM teams to position Invoca's value in RFP resulting in a 7 figure save of the largest clients.
- Generated new business by closing deals with high-profile enterprise companies: Coca Cola, Allstate, US Bank, Ecolab, and UTSW.
- Created global training and presentation materials and personally certified 80% of the sales reps (25) on platform in 2022; empowered team for sales enablement and product demonstrations using a Telecom demo site to emphasize value proposition.
- Developed new Global demo environment that highlighted Avetta's linguistic and localization capabilities across several global regions for larger clients; adopted by the whole organization and used to expand Avetta's global reach.

- Increased user satisfaction and ensured successful adoption and implementation of software by coordinating efforts with the Avetta training team for both client and employee facing training.

PROMPT: Please create a list of accomplishments in the format of the accomplishments section of the resume

> NOTE: There is no need to repeat elements such as "you are a top-notch career coach" or to re-paste your resume. ChatGPT will understand the questions in the context of the previous interactions. This means your prompts can be much shorter as you delve into the discussion.

Here is an example for one of my students:

ChatGPT RESPONSE

- *Efficient Data Migration Implementation:* Successfully managed the data migration of 250 legacy American Airlines planes into a single maintenance system, exceeding the targeted completion date by more than six months.
- *Cost-Saving AV Setup Process:* Developed an innovative approach to utilize surplus computer equipment for AV setup instead of purchasing new equipment, resulting in significant cost savings of $20,000.
- *User Error Messaging Project:* Spearheaded a project to improve user error messaging, which helped expand the company's sales by 20%.
- *Compliance Process Creation:* Designed and introduced a new compliance process for new team members covering policies, procedures, and regulatory requirements within the Technical Operations department.

- *Onboarding Process Improvement:* Overhauled the onboarding process for new hires, reducing the time to gain access to all documentation by 25%.
- *Domestic Reservation Website Enhancements:* Enhanced American Airlines' domestic reservation website products, resulting in revenue growth of more than 15%.
- *Stakeholder Matrix Initiative:* Initiated a Stakeholder Matrix that streamlined participants' approval process for regulatory changes, ensuring 100% regulatory compliance.

Resume Section: Professional Experience/Other Sections

This section is the chronological description of your job title, duties, and accomplishments. I suggest taking what you have for each position and asking the AI to restate what you wrote. There are many times when GPT4.0 can offer some more powerful or succinct words that can improve the resume.

Here is an example:

PROMPT: Improve this job description in the Experience section:

American Airlines, Ft. Worth, TX

Sr Project Manager

- Identified and analyzed project plans and timelines for projected deliverables and outcomes to shorten
- target completion date a year for data migration of 250 legacy American Airlines planes into a single maintenance system
- Spearheaded a newly implemented system of improved scheduling and access for all Records teams resulting in exceeding our targeted completion date by more than six months
- Implemented User Acceptance Testing for the benefit of the DevOps team, including new tracking of
- Supply Chain CR escalation process

ChatGPT RESPONSE:

American Airlines, Ft. Worth, TX

Sr Project Manager

- *Devised and evaluated project plans, identifying strategic opportunities to expedite deliverables, which notably accelerated the completion timeline of a data migration project by a year. This project involved the consolidation of data from 565 legacy American Airlines planes into a single maintenance system.*
- *Championed the implementation of a new system that enhanced scheduling and access for the Records team, significantly surpassing the initial target completion date by over 6 months.*
- *Directed the execution of User Acceptance Testing protocols, tailored to aid the DevOps team. This also included the development of a new tracking system for Supply Chain CR escalation processes, contributing to increased efficiency in operations.*

On other resume sections, such as Education and Languages, these data points are straightforward and do not need AI help.

Using AI to Improve your LinkedIn profile

The importance of a well-constructed LinkedIn profile cannot be overstated in today's professional climate. It acts as an online extension of your resume and allows potential employers and recruiters to gain insight into your career aspirations, skills, and accomplishments. While an in-depth guide on building an engaging LinkedIn profile is beyond the purview of this book, **engaging a professional service for a comprehensive review of your social media presence can be a worthy investment.**

Here are some fundamental tips for leveraging LinkedIn to its full potential:

- **Network Expansion:** Aim to grow your network to over 500 connections. Reach out to people you know – from high school friends, college classmates, and former colleagues. Endeavor to continuously add valuable contacts to your network.
- **Professional Profile Picture**: A professionally shot picture can make a strong first impression. Make sure your profile picture is clear, high-quality, and presents you in a professional light.
- **Customized Background Image:** Utilize the background image on your profile to communicate a piece of your personal story or to highlight your professional focus. It should resonate with your brand and what you can bring to potential employers.
- **Powerful Headline**: Your headline should encapsulate your strongest competencies and the value you can provide to a potential employer. Make it brief, compelling, and relevant to your career goals.
- **Engaging 'About' Section**: The 'About' section provides an opportunity to elaborate on your skills, experiences, and professional aspirations. Craft a narrative that effectively communicates your unique value proposition.
- **Listing Relevant Skills:** Populate your 'Skills' section with abilities that align with your career goals. The more relevant skills you

list, the better the chance of being discovered by recruiters. Ask colleagues and friends to endorse your skills.

- **Garner Recommendations:** Aim to secure at least 3 recommendations from colleagues or other professionals who can vouch for your abilities. Recommendations are often reciprocal, so be ready to return the favor.

> Suggestion: First, professionally review and update your resume, then improve your LinkedIn profile. You can copy and paste your professional experience information from your updated resume into the Experience section of LinkedIn. Have someone with social media experience review the final changes before changing your profile.

Here are the sections of LinkedIn where AI can help you improve your LinkedIn content

LinkedIn Section: Headline

This is the most important section of your profile, where you introduce your skills and qualifications to a larger audience. This appears on all search results. While a LinkedIn expert can help you with the final wording, ChatGPT can aid you in creating your initial headline verbiage by analyzing your resume content.

Here is an example from my resume prior to starting this company:

PROMPT: Write ten distinct phrases that can be used for my LinkedIn profile headline section

ChatGPT RESPONSE:

1. "Pre-Sales Maestro: Turning AI & Machine Learning Tech into Business Success Stories"

2. "Versatile Sales Engineer: Bridging the Gap between Tech & Business in Diverse Industries"

3. "Senior Sales Leader: Transforming Business Goals into Realities with AI & Machine Learning"

4. "AI and Machine Learning Advocate: Fueling Global Business Growth through Innovative Sales Strategies"

5. "Solution-Driven Sales Engineer: Expanding Business Horizons across Healthcare, Finance, Energy, and More"

6. "Cross-Functional Sales Expert: Catalyzing Business Growth with AI, ML, and Integrative Tech"

7. "Global Sales Pioneer: Harnessing AI & Machine Learning to Drive Efficiency & Revenue"

8. "Senior Sales Engineer: Translating Advanced Tech into Profitable Solutions for Diverse Markets"

9. "Innovative Sales Leader: Transforming Businesses with AI, Machine Learning, and Cross-Functional Collaboration"

10. "AI & Machine Learning Enthusiast: Fueling Business Growth in Finance, Healthcare, Energy, and Beyond"

LinkedIn Section: About Me

Writing a compelling 'About' section on your LinkedIn profile can significantly enhance your visibility and attractiveness to potential employers, clients, or partners. LinkedIn allows you to use up to 2,600 characters in your 'About' section, so use this space wisely to create a well-rounded picture of yourself. Here are some best practices to consider:

Professional Tone: Maintain a professional yet approachable tone throughout. You can write in the first person to make your summary more personal and engaging.

Start Strong: Capture attention with a strong opening line. Clearly state your current role and how your experience or skills bring value.

Define Your Role: Briefly describe what you do and how it aligns with your industry or professional interests. If you've held various roles, highlight a common thread that ties your career journey together.

Showcase Achievements: Don't shy away from mentioning specific accomplishments, awards, or recognitions you've received. Use quantifiable results to emphasize these achievements where possible.

Add Skills & Specialties: Highlight your key skills and areas of expertise. This gives readers a quick overview of your competencies and improves your visibility in LinkedIn searches.

Personalize: While keeping a professional tone is important, it's also beneficial to include something personal or unique about yourself. This can help make your profile more relatable and memorable.

Call to Action: Consider adding a call to action at the end of your summary. This could be inviting viewers to connect with you, directing them to your website, or encouraging them to reach out for business inquiries.

Use Keywords: Incorporate relevant keywords and phrases that align with your industry and role. This will help optimize your profile for LinkedIn's search algorithm, making it easier for recruiters or potential connections to find you.

Stay Updated: Keep your 'About' section updated with your latest experiences, achievements, and professional interests. This helps maintain the relevance of your profile.

Here is a prompt you can use to get the section started - Note: You will need to edit answers in the section. AI can offer ideas but do not cut and paste and consider the output as your final draft.

PROMPT: Based on my resume, please write my LinkedIn about section with a focus on a professional tone, starting strong, defining my role, showcasing achievements and skills and incorporating keywords

LinkedIn Section: Experience

You can update this section by copying and pasting the results from your updated resume. This is why we suggest first completing your resume and using that information to fill out your LinkedIn profile. Some LinkedIn experts suggest you can add more details to the job description since it doesn't have the same two-page restriction as a resume. Best practices vary case by case.

LinkedIn Section: Skills

As a user, you are allowed to add 50 skills. LinkedIn also allows you to associate specific skills with jobs that you have held in the past. Therefore, identifying your top skills (especially the ones mentioned in the job descriptions you are applying for) is important for highlighting your capabilities.

PROMPT: Please suggest 20 skills I can add to the LinkedIn skills section based on my resume.

Using AI to become an expert in your role and industry

Whether you are trying to get a job in the industry you already have experience with or are looking to transition to a new industry, ChatGPT can help you.

By now, you should have identified your target industry and desired role. The more narrow you can make, the better information you will receive from ChatGPT. Here is an example:

I could ask ChatGPT about becoming a Customer Service Manager, and it will give some good information in a general sense. If I clarify that I am targeting a Customer Service Manager role for a technology company with less than 500 employees, it will give more specific, distinct output.

PROMPT: You are an SVP of Customer Success at a 450-employee, growing software company. Please outline the top 5 priorities you have as a leader.

ChatGPT RESPONSE:

- **Customer Satisfaction and Retention:** *The primary goal of a Customer Success leader is to ensure customer satisfaction and, subsequently, retention. Regularly checking in with customers to understand their needs, address their pain points, and assess their satisfaction level with our products or services should be a top priority. The implementation of customer satisfaction metrics like Net Promoter Score (NPS) can provide quantifiable benchmarks.*

- **Onboarding and Adoption:** *Efficient onboarding and adoption are key for customers to realize the value of our products. This involves establishing an effective, scalable onboarding process that gets customers up to speed quickly and maximizes their use of our software.*

- **Customer Success Team Development:** Build, train, and empower a high-performing customer success team. This involves hiring the right people, providing regular training, and establishing clear roles, responsibilities, and performance metrics. A strong team will be equipped to handle customer needs and drive growth.

- **Relationship Management and Expansion:** Building strong relationships with key customer stakeholders is crucial. Understanding their business objectives and aligning our software's capabilities with these objectives is vital for customer retention and expansion. Cultivating these relationships can lead to upselling and cross-selling opportunities, driving revenue growth.

- **Interdepartmental Collaboration:** Effective communication and collaboration with other departments like Sales, Marketing, and Product Development are crucial. This collaboration ensures that customer feedback is incorporated into product improvements, sales are aligned with customer success efforts, and marketing understands the customer's journey and success stories for testimonials and case studies. This synergy will enhance customer experience and contribute to the company's success.

Important: Learning how to dig deeper is a useful skill to learn with prompting. The first prompt asking ChatGPT to play a specific role gave useful output and showed that the program understood the parameters. Now we can ask more probing questions to understand the world of the potential decision-maker and how best to position ourselves for the opportunities that arise.

PROMPT: You have a job opening. Please describe 5 important attributes outside of the role or industry experience you seek.

ChatGPT Response

- **Adaptability:** The ability to adapt to new circumstances, technologies, and procedures is crucial, especially in a rapidly evolving industry like software. A candidate who is open to learning new skills, and can manage changes in environment or work requirements, is highly desirable.

- **Communication Skills:** Strong verbal and written communication skills are essential in any role. These abilities are critical for collaborating with team members, liaising with other departments, and interacting with customers or clients.

- **Problem-Solving Skills:** I look for candidates who demonstrate a strong ability to analyze problems, think creatively, and devise effective solutions. This indicates that they can handle challenges and unexpected hurdles with independence and resourcefulness.

- **Teamwork and Collaboration:** While individual competencies are important, being a team player is crucial in a collaborative work environment. Candidates should show they can work well with others, respect diverse viewpoints, and contribute positively to the team dynamic.

- **Initiative and Proactiveness:** Candidates who are self-starters, eager to take on new tasks or projects, and actively seek out opportunities for improvement and learning are invaluable. This demonstrates a level of motivation and ownership that can drive both their personal growth and the progress of the organization.

While technical skills and industry knowledge are crucial, these attributes can play a significant role in determining a candidate's potential for success within the company and the broader team.

PROMPT: Describe ten ways that AI and generative AI is changing the Customer Success strategies for mid-size technology companies.

ChatGPT RESPONSE:

- **Automated Customer Support:** Generative AI, like chatbots, can handle a large volume of common customer queries, freeing up human agents to focus on more complex issues. This improves efficiency and reduces the response time for customer inquiries.

- **Personalized Experience:** AI can analyze customer data to understand individual preferences and behaviors, allowing companies to provide personalized experiences, recommendations, and support, enhancing customer satisfaction and retention.

- **Predictive Analytics:** AI can predict customer behavior by analyzing historical data. It can identify potential churn risk and suggest proactive measures to retain those customers, thus improving customer success.

- **Customer Segmentation:** AI can process vast amounts of customer data to create detailed customer segments. This helps in providing targeted, relevant communications and services to different customer groups.

- **Improved Onboarding:** Generative AI can be used to create customized onboarding materials and training modules based on the user's needs, helping customers get the most out of the product or service from the start.

- **Sentiment Analysis:** AI can analyze customer feedback and social media mentions to gauge sentiment about a product, brand, or service. This can guide customer success strategies and highlight areas needing improvement.

- **Self-Service Portals:** AI can power intelligent search and knowledge bases in self-service portals, making it easier for customers to find the information they need without contacting customer service.

- **Real-time Updates and Reporting:** *AI systems can generate real-time updates and reports on various aspects of customer success, such as usage stats, support ticket trends, and more, enabling a more responsive approach.*

- **Advanced Issue Resolution:** *Generative AI models can analyze and generate step-by-step solutions to complex product issues or queries, helping Customer Success teams provide quicker and more effective support.*

- **Customer Journey Mapping:** *AI can help track and analyze the customer journey at every touchpoint, providing insights into how the customer interacts with the product or service over time. This understanding can lead to strategies designed to improve customer satisfaction and loyalty.*

Chapter 3: Leveraging AI for Outreach

Job Matching

One of the most frequently asked questions is how to leverage AI for job matching. This is the one area where ChatGPT is least helpful. Having a strong resume and an updated LinkedIn profile gives you the best chance of attracting hiring managers and recruiters for available roles based on the skills you project.

Platforms like Indeed, Glassdoor, and LinkedIn have job postings and allow you to filter your search based on your preferences. Additionally, visit specific company websites where you are interested in working and check their career page for job openings. LinkedIn offers job matching features, like "Jobs you may be interested in" and "Alerts for jobs in your field."

Networking and industry events are also great sources for job leads.

As of this writing, ChatGPT's knowledge of data ends in September 2021. Therefore, new job openings, data about startup companies, or larger companies' performance are unavailable.

It's up to the job seeker to leverage all available resources to identify current job openings and get as detailed a job description as possible. In addition, I recommend applying directly on the hiring company's website, as opposed to a 3rd party like LinkedIn.

In situations where you want to research the background of prospective companies, databases like Crunch base and Google searches can give you additional information.

The techniques I'll teach for the rest of the book are designed to help job seekers understand their potential and how best to project their skills and capabilities. This can only be done on a subset of opportunities, and therefore job seekers should set clear criteria before sending out resumes. Here are some suggestions for how you can manually filter jobs before running an AI analysis:

- **Job Title:** If you're clear on the role you're interested in, use the specific job title to filter positions. For example, if you're seeking a 'Senior Sales Engineer role, use this as a filter criterion.
- **Industry or Sector:** If you prefer a certain industry, like software development, healthcare, finance, retail, etc., use that as a filter.
- **Company Size:** Some people prefer working in a start-up environment, while others prefer established companies. Filtering jobs based on the company's size can help align your job search with your preference.
- **Company Culture:** Though a bit subjective, some job postings give an insight into the company's culture. Look for keywords that align with your preferences, such as "innovative," "collaborative," or "results-oriented."
- **Location:** This is an important filter, especially if you're not open to relocating. You can filter jobs based on the city, state, or even country where you want to work.
- **Experience Level:** Most job postings will specify whether the role is for entry-level, mid-level, or senior-level professionals. Filtering according to your career stage can help you find suitable roles.
- **Salary Range:** Some job postings may mention the salary range. If they do, you can use this information to filter jobs based on your salary expectations.
- **Type of Employment:** You can filter jobs based on the type of employment, such as full-time, part-time, contract-based, remote, or in-person.
- **Skills Required:** If you have specific skills that you want to use in your next job, look for job postings that list these skills in their requirements or job description.
- **Keywords:** Use keywords that are relevant to the kind of job role you're seeking. For example, if you're interested in roles that require

leadership, you can filter for job descriptions that include the word "leadership."

Once you have identified one or more opportunities, let's proceed to the next section.

Applying to Jobs

Using AI for understanding compatibility

Congratulations! You have made it this far and have

- Determined your career path
- Updated your resume
- Optimized your LinkedIn profile
- Identified open jobs either via recruiters, networking or on job sites

Now you can begin targeted outreach; you have the necessary tools to evaluate whether an opportunity is a good fit, and you can optimize your chances.

While a job seeker can look at a job description and other demographic elements related to the company and opportunity to quickly determine if a job is in the ballpark of being a good fit, it would take 10-15 minutes to dig deep in the job description and see how good of a fit the job might be.t.

ChatGPT can give you a compatibility rating quickly and summarize the important points. Some job search tools match based on keywords that are not as accurate as AI. For example, a job description might call for C++, while the job seeker might know Java. The AI algorithm can estimate that while the job seeker does not have the exact skill, he or she has a similar skill and would likely be a better fit than someone who has no prior programming background.

Regardless of your tool, compatibility estimates are just that—estimates. In reality, many job descriptions are poorly written or inaccurately describe the job that needs to be accomplished. In addition, a job seeker might have a compatibility rating of 85% with the hiring manager but a 75% compatibility with an executive-level director. Finally, there is a level of randomness in the evaluation, and if you rerun the evaluation, you might find a variation of 5-10%.

How to interpret the ChatGPT Output:?

90-100% - You are a top candidate, which is necessary in a tight job market. Be aware you are not labeled as overqualified. Suppose a job posting has many applicants for a position. In that case, AI screening tools allow companies to be more picky and only let people with very high levels of compatibility pass the next steps.

80-89% - This is a job where your qualifications are close to the company's needs. Suppose you're one of a large pool of applicants, and you receive this rating. In that case, you might want to consider diving deep into your skill gaps to see how you can improve your resume to be more explicit in the skills or experience you have that tie into the needs defined in the job description.

70-79% - If the sector you are applying to has a tight labor market, then this is still a good opportunity. Most likely, 1-2 important skills or areas of industry experience are lacking, and if the competition for the job is high, you have a low chance of getting to the next level. If you were to be hired, there is a chance that there would be a longer onboarding process, and you might need more time to learn a skill or certification.

Any rating in this area should give you pause about whether this is a good fit or is worthy of pursuing. You can always submit your resume, but there is a high chance the screening tools will weed your resume out.

69% or Below - There is a large gap in skills, industry or experience. Following the instructions later in this section will help you identify the reasons and decide whether it makes sense to get a certification or learn a new skill while applying for jobs in this field, especially if you notice a repeating pattern.

Using ChatGPT to get Compatibility ratings

The contextual data we use is your resume and job description. Make sure to have both handy to copy and paste.

PROMPT: Based on a job description and resume, please evaluate the applicant's compatibility as a % and write a 2-paragraph explanation. Here is the job description: (Job Description pasted). Here is the Resume (Resume pasted)

ChatGPT RESPONSE (Actual Example from myself):

Compatibility: 85%

Ben Gold's resume suggests strong compatibility with the Pre-Sales Consultant / Sales Engineer role. His experience as a Senior Sales Engineer demonstrates that he has the necessary skills and experience to excel in this position, particularly in his adeptness at managing pre-sales processes and his proficiency in working with cross-functional teams. Furthermore, his experience in working with a range of verticals, including retail, indicates that he possesses the industry knowledge necessary for this role.

Gold's resume also exhibits his technical proficiency in APIs and integrations, which aligns with the Pre-Sales Consultant's need to understand the TradeBeyond product suite, particularly its supply chain management platform. His experience in inside and outside sales for enterprise software companies aligns well with the job requirement, as does his willingness to travel extensively, which would suit the 50% travel requirement. However, his absence of experience in specific systems like Salesforce.com, Zoominfo, or MS Office Suite might be an area that needs development.

Additional Prompts:

PROMPT: Evaluate the skills compatibility of the applicant as a % and write a 2 paragraph explanation.

PROMPT: Provide a skills gap analysis

PROMPT: For Skill ABC, how long would it take someone with beginner level understanding to master the skill?

PROMPT: Evaluate the Industry knowledge compatibility of the applicant as a % and write a 2 paragraph explanation.

PROMPT: Evaluate the experience compatibility of the applicant as a % and write a 2 paragraph explanation.

PROMPT: Give the top 3 strengths and top 3 weaknesses of the resume for this job description

PROMPT: Give 5 resume suggestions to make the applicant more compatible with the job description

After you receive AI feedback, you can update your resume to tailor it to the specific job and re-run the analysis.

Cover Letters in 10 Minutes or Less

ChatGPT has revolutionized the job search process, particularly in aiding job seekers craft compelling cover letters. Once a difficult task, writing a concise cover letter allowed job seekers to demonstrate their understanding of the job requirements and articulate their unique skill set. Typically, crafting a noteworthy cover letter could consume 30-45 minutes, depending on the complexity of the opportunity. Now that same task can be accomplished in less than 10 minutes.

Are Cover Letters Becoming Obsolete?

The relevance of cover letters in today's job market is a topic of ongoing debate. In instances where recruiters directly contact candidates, cover letters often become superfluous. Moreover, many job postings no longer offer an option to submit a cover letter. Additionally, the rise of AI in generating cover letters introduces a new set of potential pitfalls, particularly if executed poorly.

Best Practices for Crafting Cover Letters:

- Identify 2-3 key points that make your candidacy compelling.
- Review the output from ChatGPT, and request more succinct drafts if necessary.
- Always thoroughly edit AI-generated drafts before sending them out.

PROMPT: Here is a resume and job description. Please write a cover letter. Here is the resume: (Paste Resume). Here is the job description (Paste Job Description)

Here is an example of a very long and wordy example:

ChatGPT RESPONSE: Dear Hiring Manager,

I am writing to express my interest in the Senior Talent Brand Manager position at ABC Company, as advertised on your careers page. I am attracted to ABC Company's innovative approach to software development, and its commitment to providing flexible, remote work opportunities. As a marketing professional with over 10 years of experience, I believe that I could bring a valuable set of skills and perspectives to your team.

In my current role as Director of Marketing, I have been successful in driving significant increases in brand awareness and recruitment metrics. My work in creating and implementing robust marketing strategies, coupled with my deep understanding of recruitment processes, has resulted in successful campaigns that have effectively attracted top talent to our company.

Moreover, my extensive experience with software platforms like Salesforce and Success Factors will allow me to quickly get up to speed with ABC Company's tools. While I have not directly used ABC Company in my past roles, I am eager to familiarize myself with it and confident in my ability to master new tools quickly.

I understand that ABC Company operates on a fully remote basis. I am excited about this prospect and confident that I can excel in this environment. My experience has involved significant collaboration with remote and international teams, and I am well-versed in employing strategies to maintain effective communication and teamwork in such contexts.

Lastly, one of my passions is building and nurturing inclusive cultures. In my previous roles, I have successfully launched initiatives that celebrate diversity and create a sense of belonging for all employees. At ABC Company, I would be committed to upholding these values and further enhancing the company's reputation as an inclusive, forward-thinking workplace.

I am excited about the possibility of contributing to ABC Company's mission of providing efficient and flexible solutions to software developers around the world. I look forward to the opportunity to discuss my qualifications in more detail.

Thank you for considering my application.

Sincerely,

The above cover letter would be thrown out immediately because it is long, wordy and written by AI. The next prompt can fix this:

PROMPT: Reduce by 3 paragraphs. Make this concise.

ChatGPT RESPONSE:

Dear Hiring Manager,

I am writing to apply for the Senior Talent Brand Manager position at ABC Company. With over 10 years of marketing experience, specifically in driving brand awareness and recruitment, I believe I can bring a unique and beneficial perspective to your team.

In my current role as Director of Marketing, I have successfully implemented marketing strategies that have attracted top talent. I have extensive experience with Salesforce and Success Factors, and I am confident in my ability to quickly adapt to and master ABC Company's tools.

ABC Company's commitment to remote work aligns with my proven track record of collaborating effectively with international teams. Furthermore, I have a passion for fostering inclusive cultures, and I am eager to contribute to ABC Company's mission of providing efficient and flexible solutions to software developers around the world.

Thank you for considering my application.

Sincerely,

Please note that this response is much more concise and to the point. When the answer is not what you want, adding a prompt with. clearer language is generally better Examples include:

- Reduce
- Expand
- Dive Deeper
- Make formal
- Make less formal
- Focus on the two most important skill matches

Using AI for Networking

Networking is an important skill for uncovering the hidden job market, and the standard networking vehicle for most job seekers is LinkedIn. Updating your LinkedIn profile ensures you'll have as many quality connections as possible, and the larger your network is, the better chance a person you are reaching out to is connected to someone you know.

This section's advice focuses on how to engage with people on LinkedIn . Still, the techniques for building insights and engagement can work anywhere you interact with other professionals.

When you're networking, have clear goals as to why you want to connect. Reasons for networking are many:

- **Informational Interviews:** To gain in-depth insights about a particular role, industry, or company from a professional in the field.
- **Learn about a Company:** To understand the culture, values, mission, and work environment of a potential employer.
- **Learn about an Industry:** To grasp trends, challenges, and opportunities within a specific industry.
- **Uncover Job Openings:** To discover job opportunities that may not be widely advertised.

- **Expand Professional Network:** To meet professionals in your field or industry who can provide advice, mentorship, or job leads.
- **Personal Branding:** To articulate and refine your professional narrative and career goals.
- **Career Guidance:** To gain advice on career paths, necessary skills, and advancement strategies from seasoned professionals.
- **Mentorship:** To find a mentor who can guide you through your job search and career development.
- **Gather Industry Jargon:** To understand specific industry language, which can be used in resumes, cover letters, and interviews.

Once you have decided why you want to network, there are many ways to begin outreach:

#1 Create a prioritized list based on networking goals. Examples include:

- I would like to get 5 informational interviews from people who are product managers to learn about the role.
- I want to work at Salesforce and learn about the company from current and former employees.
- I want to switch from sales to project management and speak to people with similar career choices.
- I am seeking a sales leadership role for a SaaS company with 50-100 employees. Any CRO or SVP of sales in this company would be great to reach out to.

#2 Find a database you want to leverage to identify and narrow potential target companies based on location, size, industry, and performance.

#3 Use LinkedIn to determine how best to filter your results:

Figure 6: LinkedIn Search[6]

In the search bar, you can search by:

- People
- Posts
- Companies
- Groups
- Jobs
- Products
- Services
- Events
- Courses

In addition to the search categories, you have in-depth filter options. For this networking exercise, I will drill into the filters you can use for People and Companies.

People Filter Options:

An important component is you can sort by connection levels:

1st = Direct connection

2nd = Connection of someone in common (in this case, you can evaluate who the common connection is and see if your common connection can make an introduction to the person you want to network with)

3rd = Connection distance between you and the person you want to network with

---

[6] https://www.linkedin.com/search

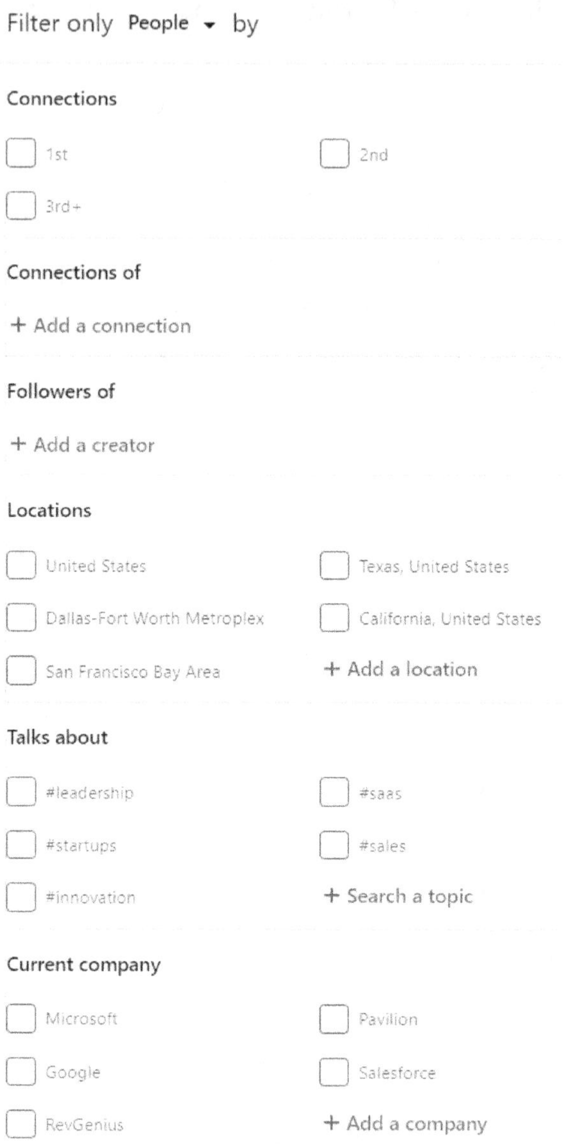

Filter only People ▼ by

Connections

☐ 1st   ☐ 2nd

☐ 3rd+

Connections of

+ Add a connection

Followers of

+ Add a creator

Locations

☐ United States   ☐ Texas, United States

☐ Dallas-Fort Worth Metroplex   ☐ California, United States

☐ San Francisco Bay Area   + Add a location

Talks about

☐ #leadership   ☐ #saas

☐ #startups   ☐ #sales

☐ #innovation   + Search a topic

Current company

☐ Microsoft   ☐ Pavilion

☐ Google   ☐ Salesforce

☐ RevGenius   + Add a company

Figure 7: LinkedIn Search Filters[7]

---

[7] https://www.linkedin.com/search

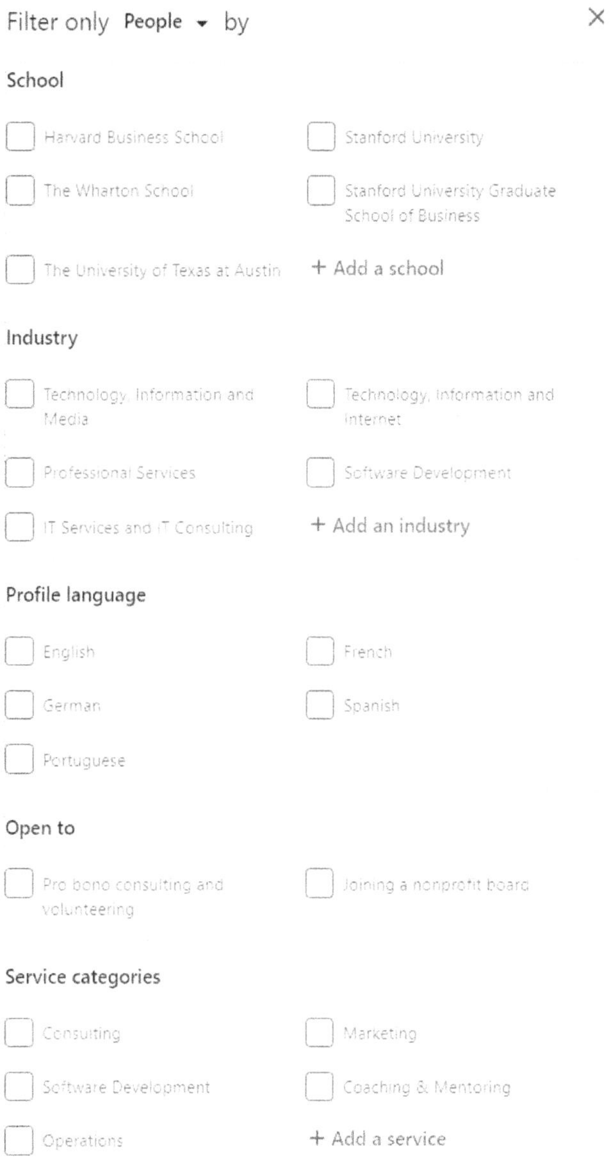

Figure 8: LinkedIn Search Filter[8]

---
[8] https://www.linkedin.com/search

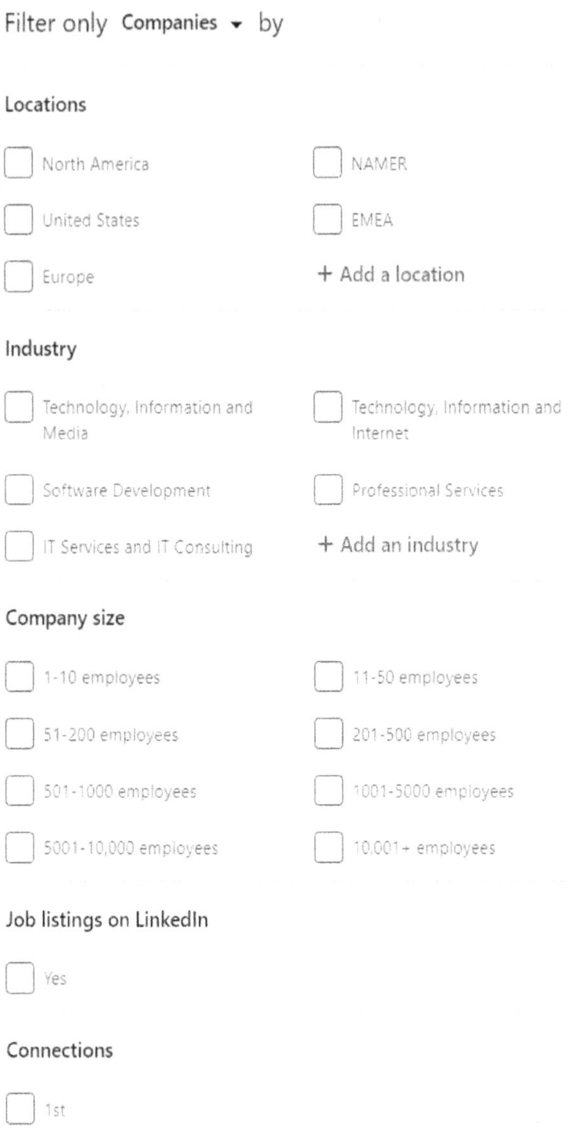

Filter only Companies ▾ by

Locations

☐ North America   ☐ NAMER

☐ United States   ☐ EMEA

☐ Europe   + Add a location

Industry

☐ Technology, Information and Media   ☐ Technology, Information and Internet

☐ Software Development   ☐ Professional Services

☐ IT Services and IT Consulting   + Add an industry

Company size

☐ 1-10 employees   ☐ 11-50 employees

☐ 51-200 employees   ☐ 201-500 employees

☐ 501-1000 employees   ☐ 1001-5000 employees

☐ 5001-10,000 employees ☐ 10,001+ employees

Job listings on LinkedIn

☐ Yes

Connections

☐ 1st

Figure 9: LinkedIn Company Search Filter[9]

---

[9] https://www.linkedin.com/search

Ways to interact with people on LinkedIn

#1 Engage with a post from someone of interest.

Approaches to effective networking on LinkedIn have matured over time. A powerful strategy to engage and garner attention is showing genuine interest in someone's posts. Engaging with their content through likes or comments increases the likelihood of their reciprocation, whether by responding to a connection request or an InMail.

ChatGPT can play a pivotal role in this process by helping you craft thoughtful and intriguing comments. It assists in demonstrating genuine interest in their narratives, significantly improving your chances of kick starting a 30-minute networking conversation.

PROMPT: Based on the content of this LinkedIn post, write 3 short examples of insightful comments that would lead to a positive engagement with the poster (Paste post)

ChatGPT RESPONSE:

1. "Thanks for sharing these insightful statistics. It's clear that remaining cautious and grounded during the pre-seed stage is crucial for founders. Overvaluing too early can indeed lead to potential issues down the line. How do you recommend founders balance this with the growing round size trends?"

2. "Interesting data points, particularly the fluctuating monthly valuations. It's fascinating to observe the MoM growth in average round sizes while maintaining reasonable valuations. I'd love to hear your thoughts on how this dynamic may evolve in Q3."

3. "Insightful post, particularly your advice for pre-seed founders. The balance between raising enough to fuel growth and avoiding overvaluation is indeed tricky. Could you expand on how these statistics may differ across various sectors?"

#2 Writing Connection Requests Where They Will Accept—300-character limit

LinkedIn allows you to communicate with a person of interest when you attempt to make a new connection. When you have identified someone you want to network with, you can use the 300-character limit to expand your network and potentially get a conversation with the target audience.

First, download the PDF profile of the person you want to network with (located below their profile picture).

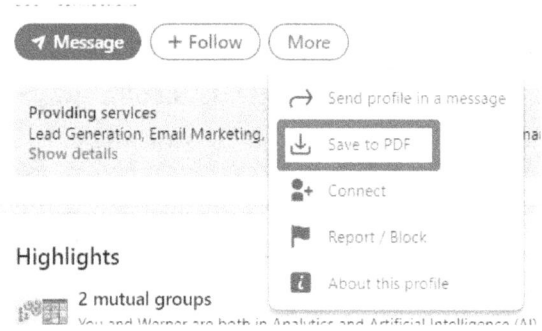

Figure 10: Saving A Profile On LinkedIn[10]

PROMPT: You are a tenured career coach. I want to send a message in the LinkedIn connection request box for an informational discussion about their background as a business owner. First, describe to me the top 3 priorities of this person. Then give me 3 examples of a LinkedIn Connection request message that incorporates my understanding of their priorities in under 300 characters. Here is their profile: (Paste Profile)

ChatGPT RESPONSE:

Top 3 Priorities for Joe Smith:

- **Business Growth:** *As evidenced by his role at Sales Performance and previous positions, Joe's primary focus seems to be on helping B2B*

---

[10] https://www.linkedin.com/

companies grow their sales. He assists with lead generation and prospecting, both crucial elements of business development.

• **Sales Expertise:** With a wealth of sales experience across different tech sectors, Joe prioritizes the application of this expertise to improve his clients' sales results, specifically within the B2B tech space.

• **Client Results:** Joe appears committed to delivering measurable results to his clients, indicating a priority on achieving client satisfaction and tangible outcomes from his sales and business development efforts.

LinkedIn Connection Request Examples:

• "Hi Joe, I'm intrigued by your wealth of experience in boosting B2B sales and would love to learn more about your approach. Could we arrange a time to discuss your business growth strategies? Best, [Your Name]"

• "Hello Joe, as an admirer of your work in lead generation and sales, I'd be keen to connect and learn more about your successful methods in B2B tech sales. Could we possibly chat further? Regards, [Your Name]"

• "Greetings Joe, I've been following your impressive work in driving sales growth and would appreciate an opportunity to connect and discuss your experiences as a business owner. Would this be possible? Thank you, [Your Name]."

Other Prompt Ideas:

PROMPT: I want to set up an informational interview to understand the best strategies to enter the digital marketing world. We are alumni of the same university. Give me 3 examples of a LinkedIn connection invite (under 300 characters) showing this person that I understand who they are.

PROMPT: This person is a LinkedIn influencer in an area I am interested in. Write me 3 examples of a compelling connection invite to show that I understand who they are. This invite aims to create a connection and get a 15-minute conversation.

Copying the GPT text to the connection invite:

Below the profile picture, click the "More" button to see the drop down options. Click "Connect."

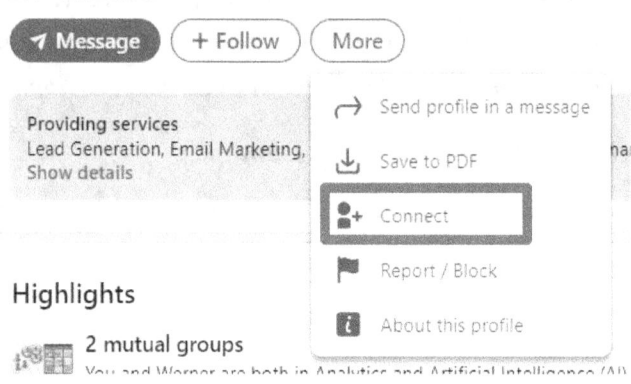

Figure 11: Connecting With People On LinkedIn[11]

A new box opens up. Click the "Add a Note" button.

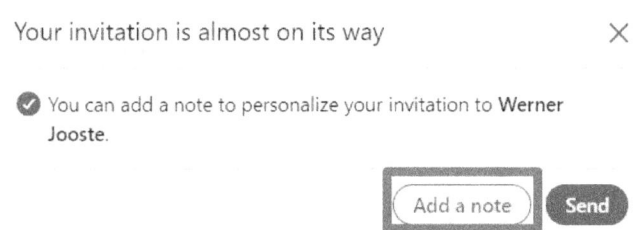

Figure 12: LinkedIn Invitation[12]

---

[11] https://www.linkedin.com/
[12] https://www.linkedin.com/

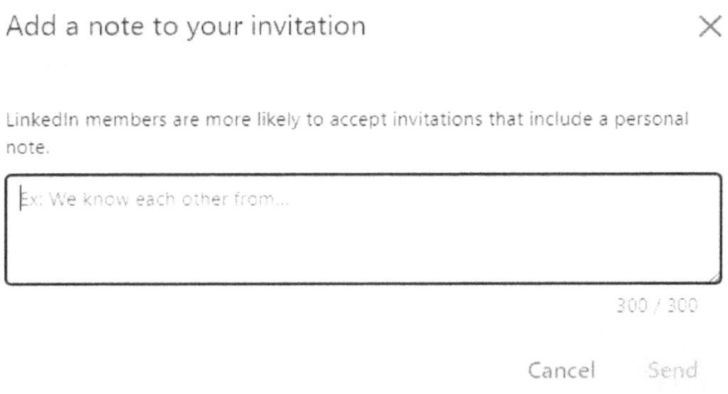

Figure 13: LinkedIn Invitation Note[13]

#3 - *Write a LinkedIn InMail or Email*

LinkedIn provides a premium feature known as InMail, which enables users to directly reach out to people they are not yet connected with. LinkedIn allows a certain number of InMails to be sent each month, depending on the subscription plan. This service can be particularly beneficial when a specific requirement calls for a more detailed message to be sent to a potential connection.

PROMPT: I am a sales leader looking to find a sales leadership position with a growing technology company. I want to write a short InMail to introduce myself. Based on my resume and the social media profile of this person, give me two examples of messaging I can use to get a response.

---

[13] https://www.linkedin.com/

PROMPT HUMOR:

Hey ChatGPT, I want to connect with you on LinkedIn. Can you help me?

ChatGPT RESPONSE:

As an artificial intelligence, I don't have a personal LinkedIn profile or any social media presence, as I'm a computer program created by OpenAI. I'm designed to provide information and assist with tasks to the best of my ability, but I don't have personal social media accounts or the ability to form personal connections. However, I'm here to help you craft an engaging LinkedIn connection request or answer any questions you have about networking on LinkedIn.

Using AI for Job Interviews

ChatGPT can help a job seeker get personalized advice for job interviews. There are many types of interviews: recruiter screening calls, hiring manager interviews, executive-level interviews, and presentation expectations. The more data you can feed AI, the more personalized your preparation response will be.

The contextual data you will need is your resume, the job description, and the LinkedIn profile of the person interviewing you. While the free versions of GPT can give you reasonable responses, this part requires 4.0.

Below is an example of several prompts and responses received in preparation for an upcoming interview.

PROMPT: I have an upcoming interview. Based on my resume, the job description and the interviewer profile, share the top 5 items we can build rapport with because we share similar skills, backgrounds or interests. Here is my resume: (Paste Resume), Here is the job description (Paste Job Description), here is the interviewer's profile (Paste Profile)

ChatGPT RESPONSE:

Based on your resume, the job description, and the interviewer's profile, here are the top 5 areas that could help in building rapport:

- *Sales Experience:* Both you and Joe have extensive experience in sales roles. Discussing the challenges and rewards of working in sales, as well as sharing some of your strategies for lead generation, sales conversion, and client retention, could help establish a common ground. You could also discuss the various industries you've worked in and how that broad experience has enhanced your sales skills.

- *Technical Proficiency:* Your resume shows strong technical knowledge in AI, Machine Learning, and APIs. Mr. Smith has a background in industrial automation and computer software sales. Sharing insights about how technology can facilitate sales and improve customer experience can be a good conversation starter.

- **Leadership:** *You've shown leadership skills by leading collaborative efforts with cross-functional teams and mentoring sales reps. Similarly, Mr. Smith has a strong military background, implying leadership experience. Discussing leadership styles, team management experiences, and methods for empowering teams could foster mutual understanding.*
- **Client Onboarding and Transition:** *Your role in ensuring successful software adoption and implementation by coordinating efforts with the training team for client and employee training aligns with the role requirement of transitioning the client to on-boarding. Sharing your experiences and best practices in this area could be a good talking point.*
- **Global Experience:** *You've had extensive global experience and have worked with diverse teams. While Mr. Smith's profile does not explicitly mention global experience, discussing the benefits and challenges of operating in an international context could be insightful. Your language skills may also be an interesting topic.*

PROMPT: Looking at the interviewer's role in the company and job requirements in the job description, what are his top 5 priorities?

ChatGPT RESPONSE:

- **Lead Generation and Sales Execution:** As a Regional Vice President of Sales, one of Joe's top priorities is to generate leads and oversee the entire sales cycle. This includes qualifying prospects, understanding their needs, and closing the deal.
- **Team Leadership and Collaboration:** Joe needs to lead a team and collaborate effectively with various other departments within the organization to achieve the company's sales objectives.
- **Client Onboarding and Retention:** After closing the sales, Joe must ensure a smooth transition from proposal to onboarding. He needs

to establish a robust follow-up process to retain clients and promote long-term business relationships.

- **Business Growth and Expansion:** Joe's role involves strategic planning and execution to grow the business. He would be looking at market trends, competition, and customer needs to make decisions that drive growth and profitability.
- **Personal and Professional Development of Team:** As a leader, Joe is likely focused on the development of his team. He would want to ensure they are equipped with the necessary skills and resources to excel in their roles and contribute to the company's growth.

PROMPT: What are the top 5 questions I will likely be asked

ChatGPT RESPONSE:

1. **Experience with HCM Sales:** Given that the job is for a Major Accounts HCM Sales Executive, you will likely be asked about your previous experience selling Human Capital Management (HCM) products. This could be something like: "Can you describe your experience selling HCM solutions to major accounts?"

2. **Lead Generation and Conversion:** Since the job description mentions self-generating leads, qualifying prospects, and closing sales, you might be asked: "Can you describe your process for lead generation, qualification, and conversion? Can you share an instance where you successfully turned a lead into a major client?"

3. **Collaboration and Teamwork:** The interviewer may want to understand how well you work with others in a sales environment. You might be asked: "Can you share an example of a successful sales deal where you collaborated with cross-functional teams to meet the client's needs?"

4. **Handling Client Transition:** As the job involves ensuring a successful transition from proposal to onboarding, a possible question could be: "Can you describe your process for transitioning a client from the proposal stage through to successful onboarding?"

5. **Personal Sales Accomplishments:** The interviewer will want to understand your personal sales accomplishments. A typical question

might be: "Can you share your most significant accomplishment in sales? How did you achieve it, and what was the impact on your previous company?"

Digging deeper, ChatGPT can also help coach you on responses to the proposed questions. This is how I asked for feedback for question #1:

PROMPT: Please rate my answer for: "Can you describe your experience selling HCM solutions to major accounts?" My answer is: I have 10 years of experience selling and supporting HCM solutions in payroll, recruitment and technology. I understand the industry's specific needs and am ready for the challenge this position offers.

ChatGPT RESPONSE:

I would rate your answer as a 7/10. While your response highlights your years of experience and willingness to take on challenges, it lacks specific details that could make it more compelling. Interviewers usually look for concrete examples that demonstrate your skills and achievements.

Here's a suggested enhancement:

"I have over 10 years of experience selling and supporting HCM solutions in the payroll, recruitment, and benefit technology spaces. For instance, while at [Company], I was responsible for [describe a specific project or responsibility where you sold an HCM solution to a major account]. During this time, I successfully [mention a notable achievement or outcome, such as a large deal closed, a process improvement, or a client satisfaction metric]. I have a deep understanding of the industry's specific needs, and I've consistently demonstrated my ability to meet these through innovative solutions and a customer-centric approach. I'm excited about the opportunity to bring this experience and perspective to [prospective company's name]."

Remember, the more you can quantify your accomplishments (e.g., % increase in sales, number of new clients onboarded, etc.), the stronger your answer will be.

Another similar prompt example:

PROMPT: As the hiring manager, give me feedback: What parts of the answer stood out? What was missing? What did I do well, and what could I do differently?

PROMPT: What are 5 questions I should ask the interviewer based on the job description and his social media profile?

ChatGPT RESPONSE:

1. "Could you please describe the qualities that the ideal candidate for this Major Accounts HCM Sales Executive position would possess?"

2. "What is the biggest challenge your sales team is currently facing, and how can the person in this role contribute to overcoming that challenge?"

3. "Given your experience in both military and industrial automation before moving to software sales, how would you describe the company culture at ABC Company, especially with respect to teamwork and leadership?"

4. "According to the job description, the role involves 'self-generating leads.' Could you please elaborate on the tools or resources ABC Company provides to facilitate this task?"

5. "You've been with ABC Company for almost a year now. What has been the most rewarding part of your journey so far, and what opportunities do you see for someone stepping into this role?"

Remember, questions should not only reflect your understanding of the role but also your eagerness to learn more about the company and the people who work there.

PROMPT: What are the 5 areas of potential weakness I should be aware of based on my resume, the job description and the interviewer's profile?

ChatGPT RESPONSE:

- **Limited Direct HCM Experience:** While your resume shows impressive experience in pre-sales and solution engineering, it's less explicit about your hands-on experience with Human Capital Management (HCM) solutions. This could be a potential area of concern.
- **Sales Experience:** The job description places a strong emphasis on the sales aspect of the role, including lead generation, closing sales, and following up with clients. Although you have sales experience, your resume leans more toward technical sales engineering. Ensuring you highlight your ability to close deals and generate leads during your interview will be crucial.
- **Strong Business Acumen:** The job description mentions the requirement of strong business acumen. While your resume suggests you possess this quality, you may want to prepare examples demonstrating your ability to understand the business landscape and drive business growth.
- **Culture Fit:** The interviewer, Austin Blanchard, has a military background which may have influenced the leadership and team dynamics at ABC Company. This could be a potential area of concern if you haven't worked in similar environments.
- **Entrepreneurial Culture:** The company prides itself on its entrepreneurial culture, which might require a high level of autonomy, self-direction, and resourcefulness. If you're more accustomed to structured, directive environments, this might be a potential area of adjustment.

At the bottom of the LinkedIn profile is a list of companies and groups the person follows. This list is not part of the PDF download. Therefore, you must manually copy the entire list of companies this person follows. Sometimes, when the person does not follow companies or groups, this exercise is impossible.

Type in the following prompt:

PROMPT: Here are the companies that this person follows. Share 4 major insights based on this list (Paste the company list):

ChatGPT RESPONSE:

- **Broad interest in Payroll and HR companies:** The person seems to have a keen interest in payroll, HR, and workforce management companies, as they follow numerous businesses in this sector. This list includes ADP, Paychex, Paycor, Paylocity, and Paycom. This suggests they may work in HR, payroll, or a related field where understanding these services would be beneficial.

- **Interests in Tech and Automation:** Following companies like TEKsystems, Schneider Electric, Emerson, Alert Logic, and ZoomInfo implies this person is interested in technology, automation, and digitization. Whether it's for personal interest or professional development, they keep tabs on the latest trends in these fields.

- **Connections to Texas & Tennessee:** The person follows the Texas Workforce Commission and Texas A&M University, along with Middle Tennessee State University and University of Tennessee, Knoxville. This suggests a strong connection to these states, possibly due to residency, study, or professional experience.

- **Interest in Sales and Leadership:** The person follows accounts such as "The Brutal Truth about Sales & Selling Podcast," "Sales Humor," "Leadership First," and "The Daily Sales," indicating a focus on sales and leadership. This suggests that sales, leadership, and related skills might be a part of their job role or areas they wish to develop.

PROMPT: Here are the groups this person follows. Share 4 insights (Paste group list):

ChatGPT RESPONSE:

- **Geographic and Professional Connections:** The person seems to have strong ties to the San Antonio region and Texas, in general, as evidenced by their membership in groups like San Antonio Small Business, San Antonio Human Resource Management Association, Inc., San

Antonio Professionals, and Texas Young Professionals. This suggests they may live in this region or have a business or professional interest here.

• **Interests in Business and Management:** They follow the Harvard Business Review Discussion Group and multiple professional groups, indicating a keen interest in business practices and management strategies and staying updated with the latest developments in these fields.

• **Background in Technology and Manufacturing:** The person's membership in groups like Robotics Consultants, Automation, and Heat Treat Forum indicates a potential background or interest in technology, robotics, automation, and possibly manufacturing or industrial processes. This combination suggests that they may work or have worked in industries where these topics are prevalent.

• **Educational Background and Affiliations:** They are a member of the MTSU Alumni and Middle Tennessee State Alumni groups, which implies that they attended Middle Tennessee State University. They also follow the Phi Sigma Kappa Fraternity Networking Group, which could mean they were part of this fraternity during their college years, fostering an ongoing commitment to this social and professional network.

Negotiate Your Way to Salary Success

There are many resources designed to help understand best practices in negotiating salaries and benefits. General best practices:

- Find out early what the salary range is.
- If the initial range is below your requirements, let the person know, as this may not be a good fit.
- Once you have established the general parameters, focus on whether you are a good fit for the company and whether the opportunity fits you.

PROMPT: Based on this job description, please rate the salary range and benefits regarding industry standards, considering skills needed, industry and location.

ChatGPT RESPONSE:

Based on the job description, the salary range of $134,000 - $145,400 per year seems competitive for a Lead Knowledge Analyst position in the Software / SaaS Product sector, especially in a location like Boston, MA. ABC Company is a reputable firm, and the hybrid work model offers flexibility that is highly valued in today's market.

In addition, the listed benefits package is quite comprehensive and above industry standards. It includes zero-dollar health insurance premiums for ABC employees and their dependents, low co-pays for healthcare services, dental and vision coverage, gym membership reimbursements, and generous retirement contributions. Furthermore, ABC offers significant paid time off, parental leave, and family benefits, adding further value to their overall compensation package.

These elements collectively position the compensation package in a very competitive place within the industry and market, especially considering the level of experience and skills required for the role. It is, however, essential for candidates to consider their personal career goals, the cost of living in Boston, and their existing skills and experience when assessing the adequacy of this salary and benefits package.

PROMPT: What are other areas I can negotiate on this offer?

ChatGPT RESPONSE:

Here are a few areas you might consider negotiating:

- **Base Salary:** While this is often the most difficult to negotiate, especially in a large organization like BCG, you could still try if you have a strong justification, such as a higher current salary or competing offers.
- **Sign-on Bonus:** If the base salary is not negotiable, a one-time sign-on bonus could be a good alternative to increase your overall compensation package.
- **Performance Bonus:** The job description mentions a discretionary performance bonus. You might negotiate the terms, for example, the goals and measures used to determine the bonus or even the bonus percentage.
- **Equity or Stock Options:** While not mentioned in the job description, some companies offer stock options or equity to key employees. If applicable, this could be a negotiation point.
- **Professional Development and Continuing Education:** You might negotiate support for additional training, certifications, or advanced degree programs relevant to your role.
- **Relocation Expenses:** If you're not based in Boston and need to relocate, you could negotiate a relocation package.
- **Flexible Work Arrangement:** If the job is listed as a hybrid, you might negotiate the specifics, like how many days you're expected to be in the office.
- **Paid Time Off:** While ABC appears to offer generous paid time off, you could potentially negotiate for additional days, especially if you're currently receiving more at your present job.

- **Start Date:** If you require more time before commencing the new role, you might negotiate your start date.

Remember, it's crucial to approach these negotiations professionally and respectfully, focusing on a fair exchange of value between you and the employer.

The one profession ChatGPT will never replace: Comedians

Getting ChatGPT to tell jokes is difficult and sometimes cringe.

Prompt: "Why did the job seeker bring ChatGPT to the interview?"

Response: "Because when asked where they see themselves in five years, they wanted to reply with a thousand different possibilities!"

Prompt: "Why was the job seeker mad at ChatGPT after the interview?"

Response: "ChatGPT said it could do the job in nanoseconds and he was left arguing about weekends off!"

Prompt: "Why did the job seeker use ChatGPT for career advice?"

Response: "Because when he asked his mom, she just told him to wear a clean shirt!"

Law:

Prompt: "Why did the lawyer ask ChatGPT for advice?"

Response: "Because when he tried to 'pass the bar', he ended up at a pub instead!"

Medicine:

Prompt: "What did the doctor say when she used ChatGPT for patient diagnosis?"

Response: "Well, at least now I can't mispronounce the medical terms!"

Education:

Prompt: "Why did the teacher use ChatGPT for lesson planning?"

Response: "Because the last time he tried, the lesson plan turned into a summer vacation itinerary!"

Software Engineering:

Prompt: "Why did the programmer consult ChatGPT while coding?"

Response: "Because he needed someone to understand his 'code humor'!"

Marketing:

Prompt: "What did the marketer say when he used ChatGPT for creating ad copy?"

Response: "Well, that's one less 'catchy tagline' headache for me!"

Finance:

Prompt: "Why did the finance analyst start using ChatGPT to understand market trends?"

Response: "Because he thought 'bull market' was a special farmers market!"

HR:

Prompt: "How did ChatGPT help the HR manager with screening resumes?"

Response: "ChatGPT promised it wouldn't judge candidates based on their profile pictures!"

Journalism:

Prompt: "Why did the journalist ask ChatGPT for a headline suggestion?"

Response: "Because her editor said her last headline was fit for a sci-fi novel!"

Sales:

Prompt: "What did the sales executive say when he used ChatGPT to write his pitch?"

Response: "At least now I can blame the AI for the corny jokes!"

Art:

Prompt: "Why did the artist use ChatGPT for creative inspiration?"

Response: "Because the last time she found 'inspiration', it was in the fridge at 2 a.m.!"

Take Action Today and Become an AI Master to Propel Your Job Search

- Invest in GPT-4.0 ($20/month), a powerful AI tool that can broaden your understanding of AI applications.

- Allocate at least two hours daily to explore AI's influence and implications in your specific industry.

- Infuse your resume with evidence of your AI knowledge. In every cover letter and interview, demonstrate how to practically leverage this understanding.

- Seek out and subscribe to AI industry newsletters to stay informed about the latest developments and trends.

AI isn't merely a trend. It's a pivotal force reshaping both our personal and professional landscapes. Don't be left out in the cold, be bold. Employers are looking to hire people who are masters in AI.

Your 60-Day Action Plan

Ready, Set, Go!? Simply use this QR code and get my free additional booklet with step-by-step suggestions on taking the insights and information and using them today!

www.ingramcontent.com/pod-product-compliance
Lightning Source LLC
LaVergne TN
LVHW011848060526
838200LV00054B/4235